This Book
belongs to the Library of
King Edward VI's
Grammar School,
Guildford, Surrey.

Black Holes

Heather Couper and Nigel Henbest

Illustrated by Luciano Corbella

DORLING KINDERSLEY
LONDON • NEW YORK • STUTTGART

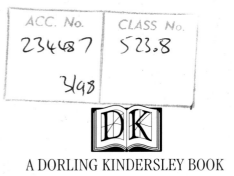

A DORLING KINDERSLEY BOOK

Editor	Jackie Wilson
Art Editor	Martyn Foote
Managing Editor	Sophie Mitchell
Managing Art Editor	Miranda Kennedy
Production	Catherine Semark
Picture researcher	Sarah Moule

First published in 1996
by Dorling Kindersley Limited
9 Henrietta Street, London WC2E 8PS
Reprinted 1996
Copyright © 1996 Dorling Kindersley Limited, London
Text Copyright © 1996 Heather Couper and Nigel Henbest
The moral right of the authors has been asserted

A CIP catalogue record for this book is available from the British Library

ISBN 0-7513-5371-X

Reproduced by Colourscan, Singapore
Printed in Italy by L.E.G.O.

Contents

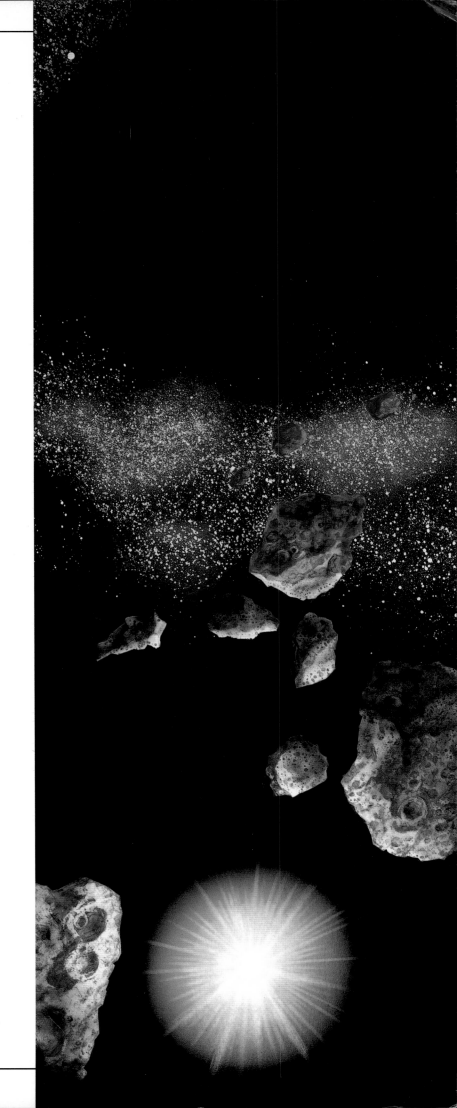

Black hole ahead

THEY ARE THE MOST MYSTERIOUS OBJECTS in the cosmos. Ravenous monsters lurking in secret places, they are also the most terrifying. They power the birth pangs of young galaxies, and may control the fate of our whole Universe. They might even be gateways to other universes, quite separate from our own. But despite their bizarre properties no one, yet, has ever seen one. Black holes. The stuff of science fiction. To astronomers, though, they are real – as real as the Sun, the Moon, and the stars – even though they are invisible. This book explores the secret world of black holes, a twilight zone at the very edge of space and time.

Black holes are aptly named. The "Black Hole of Calcutta" was a room in 18th-century India used to hold three prisoners. Once, 46 were crammed in – and 24 died. Just as in its astronomical counterpart, a large amount of matter was concentrated in a small space from which there was no escape.

A distant quasar – a young, disturbed galaxy – spews jets of hot gas into space from its glaring core. A supermassive black hole is thought to drive the violent activity.

A black hole's powerful gravity plays crazy tricks. Here, it has bent the light of a galaxy lying behind, creating a cosmic mirage. The galaxy appears closer, brighter, and split in two.

A galaxy not in line with the black hole looks undistorted. The galaxy itself may contain millions of black holes.

This black hole was created by the explosion of a massive star, a supernova. Were it not for the faint ring of light trapped by its gravity, we would not even suspect the black hole is there.

Seeing the invisible

Black holes have been around since the beginning of time – but we didn't know about them until astronomers developed new ways of looking at the Universe. Instead of using just light, today's astronomers explore space with other wavelengths. Radio waves, infrared, ultraviolet, X-rays, gamma rays – these invisible radiations have brought information of previously unknown and violent events taking place out there. Except for the very smallest, black holes emit no detectable radiation, but their gravity can have a dramatic effect on their surroundings.

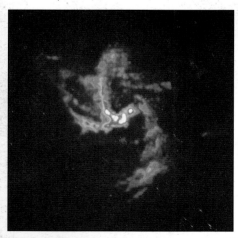

A false-colour image from a radio telescope shows a region 20 light years across at the heart of our Galaxy. It reveals a ring of hot gas circling what is probably a huge black hole.

Fragments of planets orbiting this far from the black hole can escape its clutches

From tiny to supermassive

Black holes come in all sizes. The most common ones, which future spacefarers are bound to encounter, weigh the equivalent of about 10 Suns. These holes are the remains of supernovas – the explosion of massive stars. Then there are supermassive black holes that lurk at the centres of galaxies. Created in the early days of the Universe, they have had almost 15 billion years in which to devour anything that has come too close. The biggest, weighing in at billions of Suns, drove the frenetic activity of quasars when galaxies were young. Now, these supermassive black holes lurk unseen at the heart of many apparently placid galaxies. And at the other extreme, scientists believe there are countless mini black holes the size of atoms. Created when the Universe was born, these holes have been getting steadily smaller.

Lurking among the myriad of stars in our Milky Way Galaxy, there may be millions of black holes

The second of the distorted images of a single galaxy

Debris from a planet orbiting this close will eventually be dragged into the black hole

A dazzling explosion marks the passing of a mini black hole, one that started out as heavy as a mountain but as small as an atom. This type of black hole can lose energy and end its life by exploding in a burst of radiation.

The star that became a black hole

This bleak scene shows the aftermath of a colossal supernova that wrecked a star and destroyed its family of planets. The remains of the star have become a black hole; its planets circle as fragments of debris. If they orbit close to the black hole, gravity will ultimately drag them in. Farther out, it is a different story. Although black holes have a reputation for swallowing everything, their gravitational strength drops off with distance. You can still get reasonably close and stay safe.

A birth in fire

BLACK HOLES ARE THE DARKEST THINGS in the Universe. Yet most of them start out as brilliantly shining stars. Both owe their existence to the irresistible force of gravity. Stars are created from the thinly spread atoms of dust and gas that swirl throughout space. Over billions of years, these atoms gradually clump together into dense clouds that eventually collapse under their own gravity. But the collapse does not continue indefinitely to form a black hole: other forces spring up to counteract gravity. Like air compressed in a bicycle pump – but on a far more majestic scale – the dust and gas in the cloud grow steadily hotter until a nuclear furnace switches on. The end result is a bright, shining star.

The Pleiades stars are only 60 million years old and are still clustered together in the "nest" where they were born.

Birthplace of stars

Deep inside a dark cosmic cloud, gas (mainly hydrogen) and tiny grains of dust (soot left by dying stars) start to condense, like raindrops in a storm cloud. Gravity forces the cloudlets to contract: they grow hotter and turn into "protostars".

EARLY DAYS

A protostar is surrounded by a disc of accumulating gas and dust (horizontal in this cross-section). When its core reaches 10 million °C (18 million °F), energy is released through the nuclear fusion of hydrogen into helium. A star is born.

At 5 billion years, a star like our Sun is at the midpoint of its existence

The Sun will be turning hydrogen into helium for another 5 billion years

The young star settles down to a stable existence

Jets of hot gas erupt and drive away most of the disc

Any remaining gas and dust may form planets

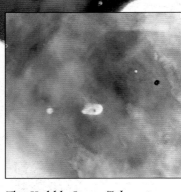

The *Hubble Space Telescope* saw these young stars, surrounded by dusty discs, in the Orion Nebula. They may be infant solar systems.

A rapidly spinning protostar splits into two

CLOSE COMPANIONS

Most stars form in twos, and stay paired, kept in place by the force of each other's gravity. Double-star systems are important in locating black holes: you cannot see a black hole, but you can see how it affects its companion.

In balance

A star like our Sun spends most of its life in an incredibly finely balanced state. Gravity is always pulling inwards, holding the star together. But at the same time, energy flowing out from the nuclear fusion reactions in the centre stops the star from collapsing. This delicate balance can last for billions of years.

Gravity wins

White dwarfs are one of gravity's victories. These shrunken corpses are the relics of stars. Once stars have run out of fuel, gravity squeezes them until the particles within cannot be packed any tighter. Most stars, including our Sun, will end up as white dwarfs.

A layer of dense gas surrounds an even denser solid core. White dwarfs have an upper weight limit: 1.4 times the Sun's mass.

The heavier a white dwarf, the smaller it is – gravity squeezes it tighter.

A matchbox of material from a white dwarf would weigh as much as an elephant!

WHITE DWARF

As the planetary nebula wafts away into space, the star's former core becomes a "white dwarf". Its nuclear reactor is dead, and it will steadily cool until it becomes a cold, black cinder.

Finally, the star puffs its outer layers into space, leaving an exposed core surrounded by a cosmic smoke ring – a "planetary nebula"

The red giant

Eventually, a star runs out of its hydrogen fuel. Gravity squeezes the star's inert core tighter, and it grows even hotter. The outer layers billow out until the star is a hundred times its former size. It is now a red giant.

The pull of gravity is exactly matched by the outflowing energy of the Sun's nuclear furnace

The core of the Sun, where its nuclear energy is produced, is over 14 million °C (25 million °F)

Gravity has a weak grip on the outer layers of the red giant, which wobble in and out

Nuclear fusion

A star is a giant nuclear fusion reactor. In its core, it converts hydrogen to helium, making the energy that keeps it shining. A star's energy reserves are huge. Every second, our Sun converts 4 million tonnes (tons) of itself into heat and light.

The nucleus of a hydrogen atom consists of a single proton – a positively charged particle. Heat and pressure within the Sun are so high that the protons bond together.

Four protons are smashed together

Two positive particles escape, converting two of the protons into neutrally charged neutrons

Two protons and two neutrons merge to form a helium nucleus

HOW THE SUN KEEPS SHINING

Einstein's famous equation ($E = mc^2$) says that any loss in mass (m) is converted into energy (E) and vice versa (c is the speed of light). A helium nucleus is 99.3 per cent as heavy as four protons. The unwanted mass is converted into energy that keeps the star shining and stops it from collapsing under the pull of gravity.

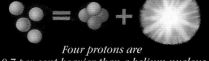

Four protons are 0.7 per cent heavier than a helium nucleus. The surplus mass is converted to energy

Gravity's ultimate triumph

ALL STARS LIVE ON BORROWED TIME. They are born of gravity and, eventually, gravity destroys them. This is spectacularly true in massive stars. A star more than 10 times as heavy as the Sun rips through its nuclear fuel at a prodigious rate – in a few million rather than a few billion years. Once a heavyweight star has exhausted its hydrogen, it has sufficiently high temperatures and pressures to fuse heavier elements. But when it tries to squeeze a core made of iron, all hell breaks loose – leading to one of the most sensational explosions the Universe can provide. A supernova explosion can spawn some bizarre descendants: a neutron star, or even a black hole.

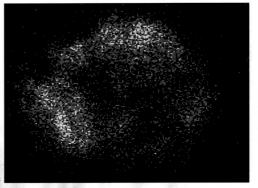

Cassiopeia A is the wreckage of a star that exploded as a supernova about 300 years ago.

Most supernovas are exploding red supergiants hundreds of times bigger than the Sun

A massive star exists in a bloated state for a hundred thousand years before gravity overwhelms it

A supernova explosion may be as brilliant as a billion Suns

Going out with a bang

A few supernovas are the result of one star in a double-star system dumping gas on the other, but most are heavyweight stars dying with a bang. Nuclear reactions have produced a core made of iron – which cannot be used as nuclear fuel. Fusing iron takes in energy rather than giving it out. The result is internal collapse: with the temperature soaring to 50 billion °C (90 billion °F), the core emits a flood of tiny energetic particles, called neutrinos, which rip the star apart.

The star's core attempts to fuse iron. To supply the energy, the star tries to…

…contract its core. The infalling matter bounces off the core and…

…powered by a flood of neutrinos, the star's outer layers are blasted into space

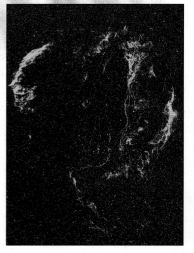

The delicate filaments of the Veil Nebula are the remains of a star explosion 20,000 years ago. From these ashes new stars will arise.

ON THE PULSE

Jocelyn Bell-Burnell and Tony Hewish stand by the radio telescope with which they discovered the first pulsar in 1967. The telescope – a huge field with 2,048 wire aerials suspended on posts – was built to study rapidly flashing sources. By chance, it found regularly repeating pulses from a neutron star.

A pulsar emits two powerful beams of radiation from its poles

Superdense lighthouses

A supernova's core collapses in just a few seconds, often producing a pulsar. These are superdense, rotating neutron stars that beam flashes of radiation – like a lighthouse – as they spin. Most pulsars, which are the size of a city such as London, spin about once a second, but the record is 642 times a second!

Pulsars only "pulse" for a million years or so. They lose energy, spin more slowly, and turn into non-pulsing neutron stars

We will never detect many pulsars: their beams are tilted at the wrong angle to sweep past Earth

If a beam sweeps past us, we detect a pulse

Magnetic pole

Rotation axis

Possible solid core

Magnetic field

Neutron fluid

Solid crust

Beam

The Crab Pulsar is the youngest neutron star we know of. It spins 30 times a second. These images capture it in its "off" (left) and "on" (right) states – "on" when we are in the beam, and "off" when we are not.

SUPERTANKER IN A PINHEAD

The material in a pulsar is much more compressed than in a white dwarf. Gravity squeezes it so tightly that a pinhead of pulsar material would weigh a million tonnes (tons)— twice as much as the world's biggest supertanker.

FULL OF NEUTRONS

A pulsar is the ultimate in squashed matter. The protons and electrons in the core of the former star have been squeezed to form neutrons – particles with no electrical charge. Standing shoulder to shoulder, the neutrons hold up the pulsar against the force of gravity. This compressed neutron star has a magnetic field about a trillion times more powerful than the Earth's. Its magnetic poles squirt dazzling beams of radiation into space.

Black out

Sometimes the relic left after a supernova explosion is too heavy to become a pulsar. If it weighs more than three Suns, not even the superdense neutrons can hold it up against the force of gravity. The object collapses even further to become a black hole.

THE SCALE OF THINGS

A star can take a lot of squeezing. When it becomes a white dwarf, a star like our Sun (1.4 million km/870,000 miles across) packs down to the size of Earth (12,000 km/7,500 miles across). A neutron star, weighing in at 1.5 Suns, is only 25 km (15 miles) across – about the size of Manhattan Island. A black hole may be just a few kilometres in diameter.

A segment of the Sun compared to a white dwarf…

…and part of a white dwarf compared to a neutron star…

…and part of a neutron star compared to a black hole

Discovery of black holes

IN 1970, AMERICAN SCIENTISTS launched a new satellite, *Uhuru*, into orbit. Its job was to track down objects emitting powerful X-rays: energetic radiation that is a sure sign of violent activity in the cosmos. *Uhuru* discovered hundreds of new X-ray sources. In many cases, the source was a compact neutron star ripping gas off a companion star. But Cygnus X-1 was different. At the same position as this X-ray source is a huge, hot blue star, about 30 times more massive than the Sun. This star is being dragged around by an unseen object weighing as much as 10 Suns – well above the limit for neutron stars. Astronomers agree that the invisible object is almost certainly a black hole, the first of several that have now been detected.

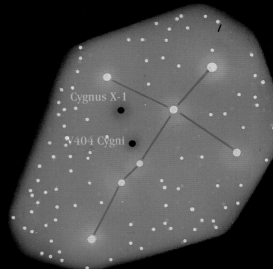

TWO IN ONE

The constellation of Cygnus (the swan) contains two probable black holes. In Cygnus X-1, the hole weighs 10 Suns, while the newly discovered V404 Cygni weighs 12 Suns.

The streamer hits the gas orbiting the black hole, creating a bright "hot spot"

This X-ray image of Cygnus X-1 was taken by the satellite *Rosat*. The X-rays are coming from superheated gas circling a black hole about 6,000 light years from Earth.

Swallowing a supergiant

Cygnus X-1 and its supergiant companion started life as a double-star system. In close-up now, the pair would be an awesome sight, with the tiny black hole – no more than 30 km (20 miles) across – relentlessly tearing gas from its companion. The gas pours towards the black hole, forming a swirling vortex called an accretion disc. As the gas falls, it travels faster and faster until it is moving close to the speed of light. Friction makes the speeding gas extremely hot, and the accretion disc glares brilliantly. Close to the hole, the gas becomes so hot that it emits X-rays.

Uhuru

Launched from Kenya on the 7th anniversary of the country's independence, *Uhuru* is the Swahili word for freedom. It made the first complete survey of the X-ray sky. X-rays cannot penetrate the Earth's atmosphere, and earlier studies had been limited to brief observations during rocket flights. *Uhuru* detected 339 sources of X-ray signals, including Cygnus X-1. The sources of the X-rays are gas heated to a hundred million degrees or higher.

The detectors on *Uhuru* could give only approximate locations for X-ray sources. Today's satellites carry sophisticated telescopes.

Black hole

As the star approaches us, the light waves bunch together. The star appears to give out light of a shorter (bluer) wavelength

As the star moves away from us, its wave fronts spread out – so we pick up light of a longer (redder) wavelength

HOW TO WEIGH A BLACK HOLE

The black hole's gravity whirls its companion star around – the stronger the gravity, the faster the orbit. If you split the light of the star into its separate colours, you can "weigh" the black hole. The amount the light changes as it approaches and moves away – known as the "Doppler shift" – reveals the speed of the orbit, and hence the mass of the black hole.

Gas is torn away from the supergiant star in the Cygnus X-1 system by the black hole's powerful gravity

The gas forms a long streamer, travelling faster the closer it gets to the black hole

X-ray pulsars

Uhuru detected many X-ray sources where gas was being snatched from a star and dumped onto a companion. Most of the snatchers were not heavy enough to be black holes and several were pulsing. They were X-ray pulsars emitting beams of radiation as gas fell onto their magnetic poles.

The gas creates a whirling vortex – the accretion disc

Up close to the black hole, the gas is hot enough to emit X-rays before it disappears forever

BLACK HOLES AND THEIR COMPANIONS

When a black hole and star are part of a double-star system, the mass of the two bodies dictates how they orbit around each other. In V404 Cygni, the hole is heavier than the star, and the star is swung around it. The centre of balance lies almost in the hole. In matched pairs like LMC X-3, the balance point lies in the middle. In the Cygnus X-1 system, the star is heavier and the centre of balance lies inside it. As a result, the super-giant merely wobbles, making it hard to be certain about the black hole's mass.

X marks the balance point

Black hole

Companion star

V404 Cygni   LMC X-3   Cygnus X-1

BEST BLACK HOLE SUSPECT

In 1989, a Japanese satellite pinpointed an outburst of celestial X-rays. British astronomers then used the William Herschel Telescope in the Canary Islands to search the same part of the sky and, in 1991, discovered a faint star, less massive than our Sun, being whirled around by an unseen object weighing 12 Suns. V404 Cygni is the most promising black hole suspect in our Galaxy because it is so massive compared with its companion.

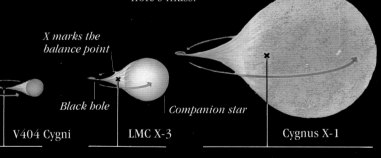

The William Herschel Telescope has a 4.2 m (13.8 ft) mirror to capture starlight.

A theory of some gravity

BLACK HOLES APPEAR TO BE SUCH A MODERN CONCEPT that it comes as a surprise to learn that they were predicted more than 200 years ago. In 1784, John Michell, an English clergyman, was pondering whether gravity had an effect on light. He suggested that some stars might be so big that light could never escape from them. A few years later – and apparently by complete coincidence – the French mathematician Pierre Simon de Laplace came to the same conclusion. At the heart of their reasoning was a theory put forward by the great physicist Isaac Newton in 1687. Newton, it is said, watched an apple fall from a tree. The reason it fell, he suggested, was due to a force of attraction called gravity. The more massive (heavier) an object, the greater was its pull of gravity. Hence the apple fell to Earth – and not the other way round.

In his garden at Woolsthorpe Manor, England, Isaac Newton ponders why an apple falls to the ground.

DISTANCE IS IMPORTANT

According to Newton, the farther apart two objects are, the weaker gravity becomes. It decreases as the square of the distance: double the distance between two objects and they feel only a quarter of the gravity. Even on Earth, an object at the top of a very tall tower weighs slightly less than at the bottom, because gravity gets weaker as you move away from the Earth's centre.

At the top of the tower the Earth pulls less strongly on the apple, and so it weighs less

At the bottom of the tower, the Earth and the apple pull on one another more strongly – so the apple weighs more

Forces at work

Newton's great leap of imagination was to realize that every object with mass has a gravitational pull. This means that the forces between an apple and the Earth and the forces that dictate the motions of distant stars are the same. At last, scientists could begin to understand why stars and planets move the way they do, and to predict how they would move in the future.

A powerful Gemini rocket is needed to launch two astronauts into space from Earth

Earth's gravity

KEEPING THE MOON IN PLACE

The Moon orbits the Earth because of the attractive force of gravity between the two. If the Earth were not there, the Moon would fly off in a straight line. But gravity is always pulling it back, and the Moon stays in orbit.

Force of gravity pulls Moon towards Earth

Moon's orbit

Straight-line path Moon would take if Earth were not there

GRAVITY INCREASES WITH MASS

Newton also found that gravity increases with mass. To break the bonds that hold you to a massive body, you must exert a strong opposite force by travelling away quickly. To leave Earth, you need to reach a velocity of 11 km/s (6.8 miles/s) to escape from gravity's pull. Any slower, and you will be pulled back to Earth. The escape velocity for the much less massive Moon is 1.8 km/s (1.1 miles/s).

The low-mass Moon has only one-sixth of the Earth's gravity. A small lunar lander is powerful enough to launch two astronauts from its surface

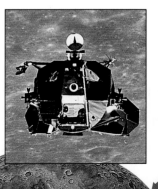

Lunar gravity

Shrinking bodies

Gravity depends on mass and distance, so you can intensify its force by shrinking a body. Imagine squeezing a spherical body of about the size and mass of the Sun. As it gets smaller, the escape velocity rises as the surface and the centre grow closer. To escape, you would need a series of successively more powerful rockets.

A CRUSH ON EARTH

If you could shrink the Earth from its present diameter of 12,756 km (7,926 miles) to the size of this model – a couple of centimetres across – its gravity would become so strong that the escape velocity would rise from 11 km/s (6.8 miles/s) to 300,000 km/s (186,000 miles/s) – the speed of light. The Earth would become a black hole.

The Earth would become a black hole if squeezed enough

ESCAPE VELOCITY

The stronger the gravitational pull of an object, the higher the escape velocity. As a dying star collapses, the escape velocity increases by the square root of the decrease in size – about 1.4 times for a star compressed to half its former diameter.

Most stars eventually collapse to become white dwarfs with an escape velocity of thousands of kilometres or miles a second

Although the object's mass stays the same, the escape velocity rises because the object is smaller and denser

To escape from a spherical body of the same size and mass as our Sun, a launch vehicle would have to travel at 620 km/s (385 miles/s) – more than 2 million km/hr (1 million miles/hr)

Crush the sphere to half its size and escape velocity goes up by 40 per cent even though its mass is the same

Squeeze it to half that size again, and escape velocity rises to 1,240 km/s (770 miles/s)

Crush the sphere to the size of the Earth, and escape velocity rises to 6,500 km/s (4000 miles/s)

When the sphere reaches the size of a neutron star, the escape velocity is over half the speed of light

Stars as big as solar systems

Instead of thinking about increasing the force of gravity by shrinking stars, John Michell reasoned the other way around. He calculated that a sphere with the same density as the Sun, but 500 times larger in size, would have an escape velocity equal to that of light and so would be invisible. In practice, no star grows this big or this massive.

A star the size of the Solar System would swallow its own light.

Trapped light rays

Black hole

Squeezed into a black hole

The natural end product of shrinking a star still further is to create a body with an escape velocity equal to the speed of light. The result is a black hole – a body with such strong gravity that even light cannot escape. Any light rays emitted from the surface would be pulled back.

A blob of interstellar gas falls towards the black hole. Far from the hole, the gas emits light in all directions

As the blob of gas gets closer to the black hole, the rays of light are bent around towards the hole

Once within the ergosphere, light rays from the gas are bent farther inward and in the direction that the hole is spinning. Until the gas passes the outer event horizon, however, some of the light it emits can still escape

Ergosphere

Static limit

...ons to predict a
...cts where gravity
...e outward. To
...m the
...the centre of
...as been
...e now
...on —
...s

...hole

...e as a
...le. A hot disc
...ion disc, may
...In the real Universe,
...cretion disc: the black
...tely black. But using
...reveal the different parts of the
...slice out of it to show what happens
...re separated by invisible boundaries: the
...nt horizon, and the inner event horizon.

In a Kerr black hole, which has spin, the singularity is elongated into a ring. It, too, is surrounded by two event horizons. Beyond the outer one is the ergosphere — a region like a cosmic whirlpool, where matter is not only dragged inward but also swirled around.

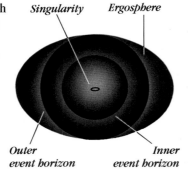

Singularity

Ergosphere

Outer event horizon

Inner event horizon

SINGULARITY

The centre of a black hole contains the entire mass of a dead star. Unable to withstand the irresistible force of gravity, the star is crushed until it becomes a point of infinite density occupying virtually no space. The point is known as the singularity. Every black hole has a singularity at its centre.

Gas in the accretion disc becomes intensely hot as it swirls around the black hole. Electrical currents flow through the accretion disc, acting like an electromagnet and generating powerful magnetic fields

Anatomy of a black hole

WITHIN A MONTH OF EINSTEIN PUBLISHING his theory of general relativity, German physicist Karl Schwarzschild discovered that the equations led to an amazing prediction. A region of space could become so distorted that it was cut off from the outside Universe. Objects could fall in, but never get out again. Today we call such a region a black hole. Einstein himself refused to believe in black holes, but for once he was wrong. At first sight, Schwarzschild's black hole looks like the one predicted by Newton's theory (see p. 17). But only Einstein's theory can explain correctly how space, light, and matter behave near a black hole. Mathematicians have even used general relativity to calculate what happens *inside* a black hole.

EVENT HORIZON

Schwarzschild used Einstein's equa[tions?] "magic circle" around massive ob[ject] is so powerful that nothing can m[ ] honour this pioneer, the distance f[rom] singularity — the massive object at a black hole — to the magic circle [is?] named the Schwarzschild radius. [We] call the magic circle an event hori[zon] because no information about eve[nts] occurring beyond this horizon can [ ] reach us. The event horizon marks the edge of the black hole. Some types of black holes have two even[t] horizons, in which case the outer one forms the magic circle.

Deeper and deeper dents

While Newton regarded objects of increasing density as having increasingly higher escape velocities, Einstein saw them as making deeper "dents" in space.

Our Sun makes a relatively shallow dent. Objects "roll" towards it at moderate speeds.

A white dwarf, being denser, dents space far more noticeably. Objects roll quickly towards it as they approach the steep slope.

A neutron star creates a dent with very steep sides. Objects rolling in reach half the speed of light.

Light rays approaching a black hole are bent around by steeply curved space

Light can escape a black hole if it gives the hole a wide berth

Rays that come closer may go into orbit around the black hole

NO ESCAPING A BLACK HOLE

A black hole makes such a deep dent that it forms a well. The sides of this gravitational well are so steep that even light cannot escape. Once anything crosses the event horizon — the boundary where the escape velocity becomes equal to the speed of light — it is trapped inside forever.

Light that comes dangerously close to the black hole is inevitably drawn in

Schwarzschild radius

Event horizon

Once inside the event horizon, light spirals in down the steep sides of the gravitational well

Inside a Kerr blac[k hole]

A massive star ends its li[fe as a] rapidly spinning black h[ole] of swirling gas, the accre[tion disc?] surround the black hole [ ] we would see only the ac[cretion disc?] hole is, naturally, compl[etely] general relativity, we ca[n] black hole, even taking [ ] inside. Different regions [ ] static limit, the outer eve[nt horizon]

THE BLACK HOLE FAMILY
All black holes have the same basic structure: an event horizon surrounding a central singularity. But there are different types of holes — stationary, spinning, and those that have an electric charge. And each has different characteristics. While one may be deadly, another may allow a journey into another universe.

The simplest is a Schwarzschild black hole. With no spin and no charge, it consists of just a singularity surrounded by an event horizon. Anything crossing the event horizon will be forced towards the singularity.

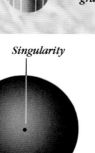

Singularity

Event horizon

In a Reissner–Nordstrøm black hole, which has charge but no spin, there are two event horizons. The region between them is a one-way zone where matter is forced to moved inward. Once inside the inner event horizon, matter is no longer sucked inward.

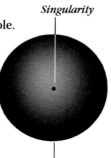

Outer event horizon *Singularit[y]*

Inner event horizon

Enter Albert Einstein

Newton's theory of gravity ruled supreme for 250 years, but it was only a partial explanation of how the Universe works. Scientists were shocked when Albert Einstein came along with his theory of relativity. In fact, Einstein proposed two theories of relativity. The "special theory" of 1905 dealt with matter, energy, and the speed of light. The "general theory" of 1915 concerned gravity. Instead of regarding gravity simply as a force, Einstein looked on it as a distortion of space itself. Where Einstein's predictions differ from Newton's, Einstein's general theory has always proved the more accurate.

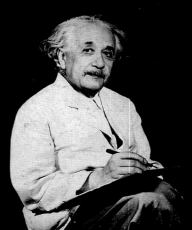

Albert Einstein was one of science's most brilliant thinkers, but he was an average student, and began his career as a patents clerk.

Real position of star

Apparent position of star

Path of light ray

The huge mass of the Sun distorts the space around it, bending the light rays as they pass

Einstein's prediction that light is bent by gravity was tested during a solar eclipse.

Einstein said that the gravitational field of an object manifests itself by distorting space. Here space is shown as a rectangular grid, warped by the presence of the Sun

Mercury

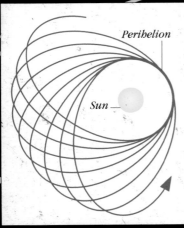

Perihelion

Sun

EINSTEIN'S RIGHT
Mercury, the closest planet to the Sun, has a markedly oval orbit which does not return to the same starting point. The point of closest approach to the Sun (perihelion) is always changing. Newton's theory of gravity cannot explain this unusual orbit, but Einstein's can. As Mercury follows the contours of warped space close to the Sun's large mass, its orbit naturally traces out a complex path in the shape of a rosette.

PASSING THE TEST
Einstein said the Sun's gravity warped space, bending light from stars passing behind it. Astronomers tested his claim in 1919 during a total solar eclipse – the only time stars close to the Sun can be seen. The apparent position of the stars shifted very slightly, exactly as predicted.

Dents in the fabric of space

Einstein thought of empty space as being like a thin rubber sheet. If you place a heavy object, such as a billiard ball, on the sheet, it makes a dent. The Sun, which is the most massive object in the Solar System, warps the space around it, making a small dent, or "gravitational well". Things moving through space follow a curved path when they meet the indentation.

Empty, three-dimensional space can be represented by a cube crossed by straight, regular grid lines

DISTORTIONS IN THREE-DIMENSIONS

To illustrate warped space, we usually draw space as having two dimensions – like the main image on this page. In reality, space is three-dimensional, and the cube on the left is how space would look without any objects in it. A massive object causes distortions, bending the grid lines that map space (below). The natural path of objects through space is not a straight line, but a curved one as they follow the humps and hollows, and "roll" towards more massive objects.

Put a massive object into space, and the regular three-dimensional structure becomes distorted

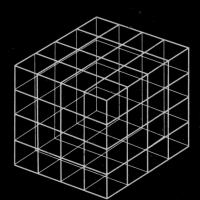

Mars

Radio signals to and from the Viking Lander on Mars are delayed by the curvature of space near the Sun – proving Einstein's theory to an accuracy of 0.001 per cent.

Venus

According to Einstein, the Sun warps the space around it. The planets whiz around like marbles in a basin, and cannot escape

Radio signals to and from Mars

GENERAL RELATIVITY IN ACTION

On Earth, we notice the effects of living in the Sun's "gravitational well". Light from distant stars is bent and radio signals from space are delayed. The difference between Einstein's and Newton's theory is hard to detect in the Sun's weak gravitational field, but it is much greater near very small, dense bodies – such as a neutron star or a black hole.

NAVIGATING BY EINSTEIN

An ocean-going yacht relies on radio signals from satellites to fix its precise position. The satellites must be programmed according to Einstein's theory of general relativity. If you used Newton's theory, the position would be adrift by 1 km (0.8 miles) every two hours.

COSMIC WHIRLPOOL

The gravitational well of a spinning black hole resembles a cosmic whirlpool — any object coming within its attraction will be swirled around as it is sucked in. Outside the static limit, a spacecraft can move where it wants. Once in the ergosphere, though, it is inexorably dragged around by the hole's spin — but it could still escape if its engines were powerful enough. Within the outer event horizon, however, it cannot get away even if its engines have infinite power.

At the outer event horizon, a rocket cannot resist the inward force of a black hole

...ce ...bole

Outer event horizon

Static limit

Ergosphere

Any kind of spinning body tends to drag space around with it, but this effect is most noticeable close to massive objects. The space around a rotating black hole is twisted as it is pulled around.

Compare this 3D structure of the space near a rotating black hole with the diagram of the structure of space near a non-rotating massive object on p. 19

Once inside the inner event horizon, the rays of light are no longer funnelled inward, but the blob of gas continues on its relentless trajectory towards the centre of the black hole

When the blob of gas reaches the centre of the black hole, its fate depends on the direction it takes. If it comes in directly over the hole's equator, it will hit the singularity and be crushed out of existence. But the path of this blob of gas is slightly tilted, so it goes through the ring in the centre of the spinning singularity

The black hole energy machine

Black holes have a fearsome reputation for sucking everything in, but it is possible, in theory, to extract vast amounts of energy from a spinning black hole. About 20 per cent of its immense energy is stored in the space being whirled around in the ergosphere.

Ergosphere

The citizens of a black hole city could even launch a spacecraft by sending it to pick up energy from the whirling ergosphere

SUPERECOLOGICAL CITIES

In the distant future, our descendants may build huge cities around spinning black holes. By harnessing the hole's spin as a source of energy, they could power a city while disposing of rubbish.

Rubbish going to ergosphere to collect energy

The trick is to aim a payload of rubbish at the ergosphere so that it orbits in the same direction as the black hole is rotating. Half of the rubbish is tipped into the black hole. The whirling gravity around the black hole acts as a sling, accelerating the other half of the rubbish and flinging it away at an incredible speed. The energy of the returning rubbish could drive a generator.

Returning rubbish brings back energy

Black holes retain only the mass, spin, and charge of objects falling in. Gravity "shaves the hair" off all other properties, including shape and chemical composition

BLACK HOLE BOMB

Imagine surrounding a spinning black hole with a spherical mirror and shining a torch through a hole in the mirror. The light bounces around inside, gaining energy each time it passes through the ergosphere. Now plug the hole. The light continues to bounce around inside the mirror, amplifying almost infinitely while the pressure builds — eventually the mirror blows apart.

A torch, a cork, and a mirror are the ingredients for a black hole bomb

The outer event horizon marks the true boundary of the black hole. Once inside it, the blob of gas and the light it emits are trapped forever. The light rays are all funnelled inward

Inner event
horizon

Outer event
horizon

From the last stable orbit to the singularity

The last stable orbit marks the closest that anything can orbit the black hole. Once within this orbit, matter is sucked into the tangerine-shaped ergosphere, a region where everything is swept around by the black hole's rotation. The outer edge of the ergosphere is known as the static limit and the inner edge as the outer event horizon. The black hole itself begins at the outer event horizon; everything passing this disappears from view for good, unable to pull itself away from the immense gravity. Inside the inner event horizon, however, the spin of the black hole sets up an opposing force to create a region of relatively normal space. Most objects will continue spiralling toward the centre, but a rocket with powerful engines could manoeuvre around (but not out of) this region. At the centre is the infinitely dense singularity, where all the matter of the original star is concentrated.

The static limit marks the point at which a rocket's engines can no longe[r] resist the sideways fo[rce] of a spinning blac[k hole]

The outer event horizon marks the true boundary of the black hole. Once inside it, the blob of gas and the light it emits are trapped forever. The light rays are all funnelled inward

BLACK HOLES HAVE NO HAIR
The person who can truly be called the "father of the black hole" is American physicist John Wheeler, who invented the name "black hole" in 1967. He also came up with the theorem "black holes have no hair". Two otherwise identical people can be distinguished by their hair colour or style. But black holes have no outwardly distinguishing characteristics. Wheeler proved that mass, spin, and charge were the only properties a black hole could possess.

"No hair" physicist John Archibald Wheeler.

HOW SMALL IS A BLACK HOLE
The more massive a black hole, the larger its event horizon is. A star ten times as heavy as the Sun becomes a black hole just 60 km (35 miles) across – about the size of Mauritius. A 20 solar mass star becomes a black hole 120 km (75 miles) in diameter – about the size of Hawaii.

The last stable orbit lies at the inner edge of the accretion disc. From here, gas tumbles into the black hole on a spiral path. Nothing can stay in orbit any closer than the last stable orbit – although a rocket can still escape by firing its engines

Naked singularities

SCIENTISTS INVESTIGATING BLACK HOLES became aware of an alarming possibility in the late 1960s. When a star collapses into a black hole, an event horizon forms and hides the singularity. But in certain situations, a black hole might form without an event horizon. Then it would be possible to see the singularity – and perhaps even to fly to it and away again. But singularities are places of infinite density, where the laws of physics break down and *anything* is possible. And without event horizons, there is nothing to protect the Universe around them: cosmic anarchy would rule. "Naked singularities" could be an irresistible target for fearless future explorers.

A stationary body shrinks to a point

A POINT OR A RING?
A singularity is gravity's final triumph – the squeezing of matter to infinite density. If the star or object being compressed is not spinning, gravity shrinks the matter symmetrically. The resulting singularity is an infinitely small point (*left*). If a spinning object is squeezed, the forces of rotation make it bulge into a doughnut shape. This shrinks, and the resulting singularity is an infinitesimally thin ring (*right*).

A spinning object shrinks to a ring

THE COSMIC CENSOR
British mathematician Roger Penrose proved in 1965 that every black hole contains a singularity. But he was so shocked by the idea of a *naked* singularity that he proposed a "cosmic censor" who would ensure that singularities are decently clothed with an event horizon. That way, the singularity stays cut off from our Universe. But Penrose has not proved that the cosmic censor exists, and other mathematicians believe that naked singularities can exist, even if only briefly.

Roger Penrose believes a "cosmic censor" forbids naked singularities.

Journey into the unknown
A spacecraft gingerly approaches a naked singularity. Formed by the collapse of a spinning star, the singularity takes the shape of a glowing ring. Inside and outside the ring, space is normal. The spacecraft can probe the singularity without being dragged in.

How to make a naked singularity

The trick is to overcome the forces of gravity that would otherwise create an event horizon. Two forces can achieve this: spin and electric charge. If a body collapsing to become a black hole is spinning very fast or has a strong electric field, the opposing force creates an inner event horizon. Increasing the spin or charge will bring the inner and outer event horizons closer together. If there is enough spin or charge, the two horizons merge and disappear completely, leaving the singularity exposed. In the real Universe, a collapsing star cannot build up enough electric charge to counteract gravity, but a very rapidly spinning star might end up as a naked singularity.

Electric forces can make your hair defy gravity…

…while rapid spin can hurl you outward

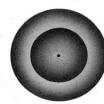

A spinning black hole has an inner and outer event horizon, with a one-way zone between the two where things can only move inward.

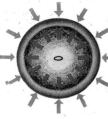

A more rapidly spinning black hole has a larger inner event horizon and a smaller outer event horizon. The one-way zone is thinner.

If the hole spins fast enough, the two horizons may merge. The one-way zone disappears, and the singularity becomes visible – and accessible.

At a ring-shaped singularity, the structure of space is kinked like a circular crease in a piece of cloth. No one can predict what will happen near there – it may even be a place where you fall off the edge of the Universe

Cosmic soapsuds

A ring-shaped singularity is not quite an infinitesimally thin line. Magnified a billion trillion trillion times, we would see the structure of space in its vicinity distorted into a "quantum foam", rather like soapsuds. Space here has no definite shape – only a set of different probable shapes.

BEATING THE COSMIC CENSOR

No one has ever seen a naked singularity, but computer simulations suggest they can form in various ways, especially when matter collapses in a very asymmetrical manner. If a long rod collapses under gravity, the simulations produce a thin, elongated naked singularity. However, it lasts only briefly before the whole mass cloaks itself in an event horizon.

With singularities being cosmic lawbreakers, much of their behaviour may be the opposite of what we expect. For instance, the most compressed thing in the Universe should also be the darkest. But physicists suggest that singularities are likely to emit radiation and shine brightly

COSMIC LAWBREAKERS

A singularity forms an edge or boundary to space, where the laws of physics break down. We cannot predict what will happen near one. It might, for example, spontaneously organize a gas cloud into a huge alien cat. If there is just one naked singularity in the Universe it could cause unpredictable chaos everywhere, even on Earth.

Falling in

BLACK HOLES ARE SO RARE that the risk of getting sucked into one is virtually zero. But what would happen if you did fall into a black hole? On the next four pages, we follow the fate of a future astronaut as she launches herself into a massive black hole (like those that may lurk in the centre of the Milky Way and other galaxies). However, things are seldom simple. Einstein's theory of general relativity reveals that the astronaut's experiences are different from what her anxious crew members – watching from the spacecraft – see her going through. It's all due to the fact that, close to a black hole, both space and time get up to amazing tricks.

Taking the plunge

Watched by her colleagues on the spacecraft, the astronaut sets off, feet-first. The spacecraft is parked at a safe distance, outside the last stable orbit. Aware that both space and time are supposed to be affected by black holes, the crew members monitor the brave astronaut's wristwatch – and also keep tabs on the light coming from their colleague and the distortion of space in her vicinity. To begin with, everything seems normal as she lets the hole's gravity pull her directly downward. Then she starts to plunge towards the hole…

SURVIVING SPAGHETTIFICATION

It would be fatal to fall into a black hole created by a dying star – one with a mass a few times that of the Sun. It warps space so severely that the astronaut falls down a very steep gravitational well. She would feel a much stronger pull on her feet than head. As she got closer to the hole, she would be stretched ever longer and thinner. Eventually, this "spaghettification" would tear her apart.

Spaghettification at the event horizon of a small black hole is equivalent to hanging from the Eiffel Tower with the population of Paris dangling from your feet

MASSIVE MEANS GENTLE

A massive black hole has a much shallower gravitational slope. If the astronaut chose a black hole of 10 million solar masses, she would feel only slight spaghettification forces. They would not be enough to kill.

Far away from the black hole, space is not distorted

Light waves coming from the astronaut are normal

The astronaut's watch and the spacecraft's clock read the same time

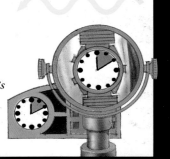

1 STARTING OUT

For the first few minutes of the astronaut's journey towards the black hole, nothing appears out of the ordinary. Her wristwatch – viewed by the crew members through a telescope – keeps the same time as the clock on the spacecraft's instrument panel; space in the vicinity (represented by the regular grid at left) is undistorted; and light coming from the astronaut is perfectly normal.

Streams of spaghettified gas falling in

Event horizon

Light is stretched to the longest red wavelengths

Near the event horizon, space is pulled out of shape

For those watching, time stops at the event horizon – the hands of her watch frozen at 12.20

3 CLOSE TO THE EVENT HORIZON
Just above the event horizon, the elongated astronaut is almost invisibl — the light has become red and dim a it loses most of its energy in the fight against gravity. Ironically, her colleagues never see her fall into the black hole. Because time runs slower and slower near the hole, she never appears to cross the event horizon, but hovers outside for infinite time.

How relativity affects time

Space has three dimensions – left-right, forward-backward, up-down. Einstein realized that time is a dimension, too – the 4th dimension. Together, space and time make up "space-time". A black hole warps not only space, but time as well.

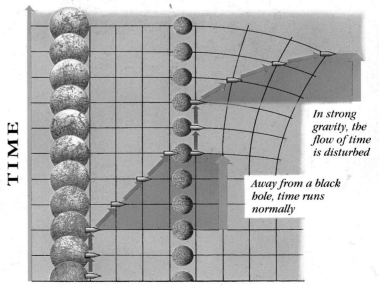

TIME

In strong gravity, the flow of time is disturbed

Away from a black hole, time runs normally

Light from the astronaut is being stretched to a longer wavelength. She begins to appear redder

Time starts to be affected by the gravity. Her watch runs more slowly than the onboard clock

Closer in, the black hole is starting to distort space

2 STRETCHING OF TIME
As the astronaut approaches the event horizon, she starts to stretch under the spaghettification forces — gravity is pulling more strongly on her feet than on her head. Although time seems to pass normally for her, the spacecraft crew can see her watch beginning to run slow. The hole's strong gravity is distorting space and time. Her colleagues also notice that she looks redder — light loses energy as it struggles against gravity.

This "space-time diagram" shows a spacecraft travelling between planets, and then through a region near a black hole (*right*). To start with, it moves in a straight line through undistorted space-time. Near the black hole, both space and time are distorted. The spacecraft follows a curved path through space-time. Effectively, time runs more slowly.

Through the black hole

THE INTREPID ASTRONAUT feels herself plunging ever faster towards the black hole, unaware that her colleagues watching from the spacecraft see her frozen in time above the event horizon. She has more important things on her mind – like the huge black hole that is looming ahead. There is no escape now. But as she crosses the event horizon, the dark void is suddenly replaced by a dizzying array of fantastic views. Space-time inside the hole is so warped that it allows glimpses of other universes. If the astronaut can carefully navigate her way through the black hole, she may be able to reach another universe.

No turning back: just above the event horizon, the astronaut sees the black hole encircled by a brilliant ring of trapped, orbiting light.

1 CHANGING VIEW
As the astronaut plunges through the black hole, the view through her helmet visor constantly changes. She sees several universes, as if through windows nested inside one another. They may have different stars and dimensions – and even unimaginable kinds of life.

Outer event horizon

One-way region – in

Inner event horizon

Polar route

Equatorial route

Bridge to another universe
Einstein and his colleague Nathan Rosen suggested that the "throat" of the black hole may open out into a mirror-image throat connected to another universe. In theory, the astronaut should be able to use this Einstein-Rosen bridge to cross to this other universe, but there are considerable dangers ahead. If the black hole is not big enough, she will be pulled apart by the spaghettification forces. If the hole is not spinning, there will be no way she can avoid hitting the infinitely dense singularity at the centre, where she will be killed. A spinning black hole with its ring-shaped singularity could provide the astronaut with a safe path. But she must navigate her way towards the singularity with great care.

To survive, you must choose your black hole carefully. It must be big, with a gradual gravitational slope. It must also spin to provide a safe way through.

Navigation is everything. If the astronaut comes in along the black hole's equator, she will hit the singularity. To get through, she must approach from one of the poles

Our own Universe, distorted by the warped space-time inside the black hole so that it appears to be in front of, rather than behind, the astronaut

2 ONLY ONE WAY TO GO

Once she reaches the inner event horizon, space is relatively normal, and the astronaut can manoeuvre freely within this region. Space-time is so warped that she may see light from several different universes, but she can reach only one of them – the one connected to her own Universe by the Einstein-Rosen bridge. Because there is no way back, she goes forward to this new universe.

The astronaut emerges through a ring of trapped light into the new universe

Ring of light, from stars and galaxies in our Universe, trapped into orbit around the black hole

Completely different universe – with its own exotic life forms

Negative universe, where antigravity operates. While gravity attracts two bodies, antigravity drives objects apart

3 ENTERING A NEW UNIVERSE

The astronaut may emerge in a universe very different from our own. It may have bizarre forms of matter and many more than four dimensions. She might not even be able to survive in this new universe. If the astronaut finds another spinning black hole, she could perhaps use this to connect to another universe. It is very unlikely, however, that she will find one that returns her to our Universe.

Singularity

This view is a snapshot of what the astronaut sees in the centre of the black hole, halfway through the Einstein-Rosen bridge

Emerging outer event horizon

Emerging inner event horizon

One-way region – out

WHITE HOLES

The exact opposite of a black hole, a white hole violently ejects matter – including our astronaut. Emerging into the new universe with her are huge quantities of matter and light that have fallen into the black hole in our Universe. The white hole is a brilliant beacon and an apparently endless font of matter and energy.

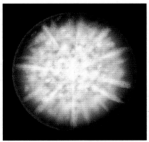

WHITE HOLE OR WHITE ELEPHANT?

Are there white holes in our Universe spewing out the miscellany of material that has fallen into a black hole in another universe? No one has ever spotted a white hole and many scientists think it would self-destruct very quickly. The ejected matter would pull itself together gravitationally and collapse into a black hole that swallows up the white hole.

Wormholes

B LACK HOLES ARE A PERILOUS WAY TO TRAVEL. Apart from the dangers of spaghettification and collisions with singularities, the tunnel that connects a black hole to another universe stays open only briefly and then collapses. But there may be an alternative, although at the moment it exists only in theory. One day, scientists may be able to turn off the fury of a black hole using antigravity – the opposite of gravity – to create a wormhole. A wormhole has two mouths, connected by a tunnel through curved space. Unlike the event horizon of a black hole, the mouth of a wormhole allows two-way traffic: you can enter and leave. And a wormhole also has the great advantage that it can connect different parts of our own Universe, providing a safe shortcut between two distant places.

BRIEF OPENINGS
A black hole provides an unstable route between our Universe and another. After a black hole forms (*left*), it briefly connects to another universe (*centre*), but the tunnel inevitably collapses (*right*). It may even close prematurely if it is disturbed, for example, by an astronaut trying to travel through.

A wormhole's mouth would look like the entrance to a non-spinning black hole. The difference is that there is no event horizon, so traffic can cross in both directions – in and out of the tunnel

One small step into a wormhole
It's the 25th century. At the Kennedy Space Center, Cape Canaveral, a NASA scientist is preparing to go to work. But he won't be using a rocket. No one has for centuries – which is why NASA's armada of launch vehicles sits gently rusting away on the tarmac, a memorial to the quaint, bygone days of rocketry. Instead, he kits himself out in his spacesuit – and enters the waiting mouth of the specially constructed Kennedy Wormhole, which is lined with antigravity material. This "one small step for a man" truly constitutes a giant leap. Stepping into the entrance, the scientist emerges in another world.

Making your own wormhole
It is one thing to keep an existing wormhole open, but there may not always be one to take you where you want to go. The answer is to create one. Make a hollow in space and then gently curve space until your destination is close to the base of the hollow. Make a small hole in the base of the hollow, and another next to your destination. Glue the edges of the holes together. You have made your own personal wormhole, and are free to travel the Universe.

HOLDING A WORMHOLE OPEN WITH ANTIGRAVITY

The tunnel formed between the two mouths of a wormhole is stable: it will not pinch off. But how do we ensure that the tunnel remains open? The trick, according to Kip Thorne, is to reinforce the walls of the tunnel with some sort of exotic material that pushes the wormhole's walls apart. Instead of having gravity, this material must exert antigravity, which forces everything away from it. Thorne believes that, one day, an extremely advanced society will develop the know-how to make an antigravity material.

Created on a car journey

Kip Thorne, an American physicist, was the first person to suggest, in 1985, that wormholes might be used for space travel. Asked by astronomer Carl Sagan to help with his novel *Contact*, Thorne solved the problem on a long car journey. Sagan planned to transport his heroine to the star Vega – 26 light years away – via a black hole. Halfway along Inter-state 5, Thorne realized that the only safe way was by wormhole.

Kip Thorne invented the wormhole, but it will take a much more advanced society than ours to build one.

Looking back into the mouth of the wormhole, you can see the light coming directly from the other end

The image of the other end of the wormhole is distorted because the light rays follow the flared mouth of the wormhole, which bends them like a lens

STRAIGHT-LINE SHORTCUT

A wormhole can provide a swift, straight-line route between two parts of our Universe, no matter how far apart they are. Since space can be curved, or folded, the length of the wormhole can stay the same, whether connecting distant or close parts of the Universe. Going by wormhole is far quicker than travelling at the speed of light to very distant parts of the Universe.

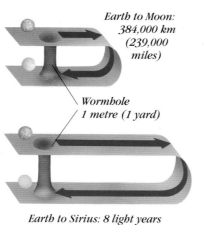

Earth to Moon: 384,000 km (239,000 miles)

Wormhole 1 metre (1 yard)

Earth to Sirius: 8 light years

One giant leap across space

The NASA scientist emerges from the wormhole into the Moon base. It has taken him no time at all to cross the 384,000 km (239,000 miles) that separate the Moon from the Earth – a journey that took the Apollo astronauts three days. Through the mouth of the wormhole, you can see the image of the rusty rockets back on Earth. That's because light also travels through the wormhole, although it is distorted by the antigravity material pushing the light beams apart. Look at the picture of the Kennedy Space Center on the opposite spread, and you'll see the corresponding image of the Moon base through the other wormhole mouth.

Time travel

ONE DAY BLACK HOLES MAY GIVE US a means of travelling through the exotic reaches of space – and possibly into other universes. They may even provide the key to making a journey through time. To be a time traveller, you need a "tamed" black hole: a wormhole. The idea of time travel through a wormhole does not seem so far-fetched when you consider that wormholes are shortcuts to very distant places in curved space (see p. 30). They take you to a remote location in almost no time at all, so it is like travelling faster than the speed of light. And Einstein's special theory of relativity says that if something is able to travel faster than light, it will move backwards through time. So wormholes may be the gateways into the past. Follow the scientist's weird experiences as he creates a time machine using a wormhole.

Bill and Ted's Excellent Adventure: two students about to flunk a history course use a phone-booth time machine to bring historical figures to the present.

Twenty years have passed on Earth, and the scientist is 40

Another 10 years, and the scientist is 50

2 OUT STEPS THE FUTURE
At age 30, the scientist finds an aged man climbing through the wormhole, followed by a gang of futuristically clad children. He is face-to-face with himself, aged 70.

On a speeding spacecraft, time moves more slowly than on Earth. When she returns, the astronaut twin is aged 35, but the twin who stayed on Earth is 70

The twin on the spacecraft speeds away from Earth at 98 per cent of the speed of light

After 5 years have lapsed on the onboard clock, the space twin turns around and speeds back to Earth

Time on Earth is passing. The scientist is now 26

TIME

The twins paradox
We are all travelling into the future as time passes, but Einstein's theory of special relativity can provide a shortcut through time. Start with a pair of twins. While one remains on Earth, her astronaut sister blasts off into space at almost the speed of light. Relativity tells us that the faster an object moves, the slower time on it appears to pass – an effect known as time dilation. When the speeding astronaut returns, she has hardly aged, while her twin on Earth is an old woman. This method cannot, however, take us back into the past.

At the start, the twins are aged 25. Both are living on Earth, and time passes at the same rate for each of them

The scientist is 20 years old when the spacecraft blasts off

Wormhole to the past

Combine the "twins paradox" with a wormhole, and you could create a time machine that allows us to travel both ways in time. Kip Thorne suggests attaching one end of a wormhole to a speeding spacecraft, while the other end stays on Earth. In this example, 50 years pass on Earth before the spacecraft returns. But on the spaceship, only 10 years have elapsed, so the wormhole connects the spacecraft with the Earth as it was 40 years earlier. By stepping onto the spacecraft and through the wormhole, future humans could travel decades back into the past.

GRANDMOTHER PARADOX

A mad scientist, intent on evil deeds, creates a wormhole. Travelling back in time through the wormhole, he finds his grandmother as a young girl – and kills her. But if he killed his grandmother, then she would not have been able to give birth to the scientist's mother. She, in turn, could not have given birth to the scientist. The scientist wouldn't exist – so how could he go back in time and murder his grandmother? Such paradoxes prompt some scientists to declare time travel must be impossible.

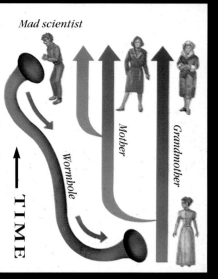

Mad scientist

Grandmother

Mother

Wormhole

TIME

The spacecraft has been away for 40 years, and the scientist turns 60

3 STEPPING INTO THE PAST

Fifty years after its launch, the spacecraft lands in the 70-year-old scientist's backyard – still with its wormhole attached. Because of special relativity, only 10 years have elapsed onboard the spaceship. This means that the wormhole's other mouth is joined to the Earth as it was 10 years after launch – 40 years ago. If the scientist steps through the spacecraft's wormhole, he can travel back through time and meet himself at the age of 30.

Young people queue for a trip through the wormhole – eager to see how things were before they were born

The spacecraft is now back on Earth. Ten years have passed on the spacecraft since it started its journey

After 5 years, as recorded on the onboard clock, the spacecraft speeds back to Earth

The spacecraft sets off at a speed close to that of light

1 STARTING FROM EARTH

An ingenious young scientist on Earth decides to construct a time machine. First, he makes a wormhole. He attaches one end of it to the Earth, and the other to an unmanned spacecraft. Next, he launches the craft so that it sets off across space at a considerable fraction of the speed of light. He has programmed the spaceship to return later on. Now, all he has to do is sit back and wait....

SHAPE OF TIME MACHINES TO COME

Scientists have dreamed up other kinds of time machines, but these are even more far-fetched than wormholes. One idea is to make an infinitely long cylinder and to spin it extremely rapidly. Another involves exotic (and as yet undiscovered) entities called "cosmic strings" – thread-like tubes of concentrated energy formed in the very early Universe. If two cosmic strings are swiftly moved past each other, they affect space-time, and might allow time travel. It is also possible that a spinning naked singularity (see p. 24) could be a time machine.

A fast-spinning, infinitely long cylinder might work as a time machine. But it would need to be made of ultradense matter to stop it from flying apart.

Exploding black holes

B LACK HOLES CAN SHINE BRIGHTLY, shrink in size, and even explode. When British physicist Stephen Hawking made this prediction in 1974, it shook the scientific world. Black holes were regarded as the ultimate sinks of the cosmos: nothing could get out, and holes could only grow bigger as they gained mass by swallowing gas and stars. Hawking's theory was an inspired leap of the imagination that combined general relativity with quantum theory – the physics of the very small. He found that energy is emitted by the gravitational field around a black hole, draining away its energy and mass. This "Hawking radiation" is negligible for most black holes, but very small ones radiate energy at a high rate until they explode violently.

The quantum view

On a very small scale, space has some peculiar properties. A pair of particles can appear out of nothing, created in a burst of energy borrowed from a gravitational field. There is always one ordinary particle, such as an electron, and its antimatter twin, such as a positron. Usually, they collide within a fraction of a second, annihilating each other and releasing the borrowed energy. But close to a black hole, one particle may be pulled inside the event horizon, leaving the other free to escape. To the outside Universe, it seems that a particle of matter has been created.

Particle-antiparticle pairs continually form and annihilate one another

Close to a black hole, the pairs start to feel the pull of gravity

One particle falls into the hole; its twin escapes. Escaping particles create a glowing halo around the black hole

THE PARTICLE AND ITS ANTIPARTICLE TWIN

Atoms – the basic units that make up everything in our Universe – are made of particles called protons, neutrons, and electrons. All subatomic particles like these have antimatter "twins", with opposite properties (including electric charge). A burst of energy can create a particle and its antiparticle: when they meet up again, they annihilate each other in an explosion of equal energy.

Physicists can create particle-antiparticle pairs in particle accelerators. Here a burst of energy produces an electron (green) and its antimatter twin, a positron (red). They spiral in opposite directions.

THE AMAZING SHRINKING BLACK HOLE

When a particle escapes from a black hole without repaying its borrowed energy, the hole forfeits this amount of energy from its gravitational field. And as Einstein's equation $E=mc^2$ says, if you lose energy, you lose mass. The black hole becomes lighter in weight and shrinks.

Near a massive black hole, space curves gently. Most of the particle-antiparticle pairs created here meet up again and annihilate each other

Two pairs have lost their partners

A QUESTION OF TIMING

Stephen Hawking spends his working life studying black holes and the origin of the Universe. Disabled by a crippling disease, Hawking cannot write or speak (he has to store complex concepts in his head). Mental arithmetic led to his prediction that black holes eventually explode, with the most massive holes having the longest lifetime. Hawking suggested that mini black holes born in the Big Bang should be exploding right now.

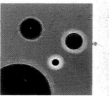

An assortment of black holes was created by the tremendous forces that existed shortly after the Big Bang that spawned our Universe

The smallest, weighing a million tonnes (tons) – about the weight of a supertanker – exploded within 10 years

Mini black holes – those weighing a billion tonnes (tons) – should be exploding now, about 15 billion years after the Big Bang

A black hole as heavy as an asteroid will live much longer than the Universe – for more than a million million million years

Hawking's mini black holes have the mass of a mountain but are the size of the nucleus of an atom

Boiling away to destruction

All black holes evaporate, but big ones boil away only very slowly. Their radiation is so dim that it is undetectable. But as the hole gradually gets smaller, the process speeds up, and eventually runs away with itself. As the hole shrinks, so the gravitational well steepens, creating more escaping particles and robbing the black hole of ever more energy and mass. The hole shrinks more and more quickly, fuelling an ever faster rate of evaporation. The surrounding halo becomes brighter and hotter. When its temperature reaches a quadrillion degrees, the black hole destroys itself in an explosion.

In the final stages, the black hole explodes in less than a millionth of a second with the energy of a billion hydrogen bombs

As the hole shrinks, it emits more particles. Its halo appears ever hotter and brighter

WHO LOSES ENERGY FASTEST

The rate at which a black hole shrinks depends on its mass. Curiously, small, low-mass black holes lose energy fastest. What is important is the curvature of space around the black hole. A small black hole has a much steeper gravitational well than a large, high-mass black hole. Just as an astronaut approaching a small hole suffers greater "spaghettification" effects (see p. 26), so the steeper well of a small hole is more effective in splitting a particle from its antiparticle twin.

In the steeply curved space near a small black hole, four pairs have lost their partners

TELLTALE SIGNS

A mini black hole explodes in a burst of gamma rays, the most energetic radiation of all. Astronomers are looking for this telltale burst of radiation, but although many objects in space produce gamma rays, none has been identified as an exploding black hole.

Downtown in the Milky Way

T HE UNIVERSE CONTAINS some immensely massive black holes – millions or billions of times heavier than the Sun. They were probably born during the early days of the Universe, when huge balls of gas accumulated and collapsed under their own gravity. Until recently, astronomers thought all of these lay a long way away, but one may live in our own home galaxy, the Milky Way. New telescopes and satellites are revealing unsuspected violence "downtown" in the Galaxy's centre, which is about 25,000 light years from the Sun. An erupting ring of dark clouds, contorted magnetic fields, racing clouds of hot gas, and a peculiar source of radio waves all point to the work of a single culprit: a supermassive black hole lurking in the heart of our own "star city".

The radio telescope at Effelsberg, in Germany, is larger than a football pitch and can be tipped to point to any part of the sky. It has revealed magnetic loops in the Galaxy's centre.

On a clear night you can see the distant stars of our Galaxy as the glowing band of the Milky Way.

Seeing through the smog

From the outskirts of our Galaxy, it is hard to see the "downtown" area because the Milky Way is thick with tiny grains of rock and soot shed by dying stars. But telescopes which can pick up infrared, radio waves, and X-rays can "see" through the smog. They show that the Galaxy has a central "hub" of old stars that date from its birth 14 billion years ago, and reveal a hotbed of activity at its heart.

COBE
The *Cosmic Background Explorer* (*COBE*) satellite was launched in 1989 to find heat radiation from the Big Bang. But it also detected infrared radiation from the central regions of the Milky Way, showing that the stars here are arranged on an oval hub, or "bar".

RADIO ACTIVITY IN SAGITTARIUS
The heart of our Galaxy lies deep inside the star clouds of Sagittarius. In the pioneering days of radio astronomy, researchers discovered two strong radio sources here – Sagittarius A and B. We now know these are clouds of hot gas associated with violent activity in the galactic centre.

The old stars in the central hub are cool stars, shining orange or red

Quiet in the suburbs

As galaxies go, the Milky Way is quite a big one. It contains around 200 billion stars, arranged in a spiral pattern 100,000 light years across. The Sun lies in the suburbs in a spiral arm about halfway out from the centre. The arms are rich in gas and dust – the raw materials of stars – so stars are still being born here. By contrast, the stars in the central hub are old and there is little activity there – apart from a tiny energetic core at the very centre.

SMOKE RING
A giant smoke ring of dark clouds, thick with dust and molecules, is rapidly expanding from a titanic explosion several million years ago. The culprit must have been a small, powerful object in the central core.

Black hole at the heart

The central core of our Galaxy – just 10 light years across – is full of weird happenings. There is no absolute proof, but they could be the work of a black hole three million times heavier than the Sun, born during the formation of our Galaxy. Only the powerful gravity of such a beast could explain the hotbed of activity.

Sagittarius B2 is the biggest dark cloud in the downtown area. It contains over 70 different types of molecules, including enough alcohol to fill the Earth with whisky!

The dense scrum in the galactic centre contains young blue stars as well as numerous old red and orange stars

Sagittarius A

Infalling gas that missed the black hole could be the raw material for the young hot stars that are shining blue in the centre

HOT GAS CLOUDS

A ring of hot gas, Sagittarius A, is swirling around several light years from the Galaxy's centre. Its speed shows that the gas is in the grip of a powerful gravitational pull – far stronger than the pull of the stars at the centre. Most likely, the extra pull comes from the gravity of a black hole.

CENTRAL RADIO SOURCE

A very small but intensely powerful radio source marks the Galaxy's exact centre. It is probably an accretion disc of superhot gas surrounding a massive black hole.

Hot gases are rushing out from the core – possibly the result of explosions in the power-packed accretion disc

The gale of gases tears away the outer layers of a red giant star, creating a long tail which makes the star look like a huge comet

Magnetic barrel

The Arc

MAGNETIC BARREL

A barrel-shaped region of strong magnetic fields surrounds the Galaxy's centre. It includes the contorted band called The Arc – narrow magnetic streamers 150 light years long but just half a light year thick. The Arc's shape indicates that there must be a powerful electric dynamo at the galactic centre. Could it be the work of a spinning black hole?

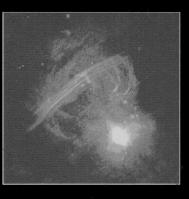

ROSAT

Rosat, an international satellite, was launched in 1990 to detect natural X-rays from space. It has discovered almost 100,000 new objects that emit X-rays, and has pinned down the position of many strange objects, including the Great Annihilator.

Great Annihilator

THE MOUSE

Radio telescopes reveal isolated patches of magnetism. This one, shaped like a mouse, was caused by a pulsar leaving a magnetic wake as it sped through space.

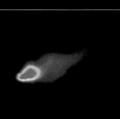

GREAT ANNIHILATOR

Just 300 light years from the Galaxy's centre, an object is spitting out two beams of anti-matter which annihilate, or destroy, ordinary matter in surrounding space. The beams are probably being generated by an accretion disc surrounding a black hole weighing in at 10 Suns. This is just one of an estimated 100 million black holes in our Galaxy produced by the deaths of massive stars.

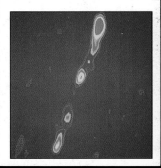

Quasars

W HEN QUASARS WERE DISCOVERED IN 1963, no one realized they were looking at objects that harboured the biggest black holes in the Universe. After all, they look just like faint stars. Astronomers soon worked out that quasars are billions of light years away, and to be visible at that distance, they must be immensely bright. In fact, they are not stars at all, but the glaring hearts of remote galaxies – star cities in such turmoil that the Milky Way's activity looks tame. The only way so much energy can be concentrated into such a small region is by the gravity exerted by a truly massive black hole. The brilliant light is an accretion disc of gas spiralling into the hole. Astronomers can weigh the black hole in a quasar by measuring the speed of orbiting stars or gas: the higher the speed, the heavier the hole. The record is a black hole weighing in at 100 billion Suns – as massive as the entire Milky Way.

Jets of charged particles – mostly electrons – shoot out from the centre of the accretion disc. The jets can be thousands of light years long

Breakneck speeds

This is the heart of a quasar – a glaring accretion disc made up of gas, torn-up stars, and dirty dust whirls at breakneck speed around a supermassive black hole, weighing in at billions of Suns. This activity powers jets that shoot into space at almost the speed of light.

GAS BOWL
A quasar's brilliant light comes from the hot gas at the centre of the accretion disc. The expansion of this gas, together with the forces of gravity and rotation, push the two faces of the disc apart, creating a bowl shape. Powerful magnetic fields speed up atomic particles in the gas and force them away as a pair of jets

ACTIVE FAMILY
Quasars have cousins in the radio galaxies and blazars: all three are often called "active galaxies". In fact, they may be one and the same. What we see depends on the angle at which we are viewing them: whether we are seeing the accretion disc and jet face-on, edge-on, or at an angle to us.

If the accretion disc is angled to us, we see a quasar: we observe the hot core of the disc and the jets are dim

The galaxy M87, imaged here by the *Hubble Space Telescope*, has a jet emerging from the vicinity of a black hole weighing 3 billion Suns.

A blazar's brilliant light comes mainly from the jet: it and the accretion disc are pointing straight toward us

In a radio galaxy, the edge of the accretion disc is facing us, and it obscures the hot, bright core. The jets may be observable by a radio telescope

The jets of an active galaxy give out strong radio waves, and are most easily observed with a radio telescope

DISCOVERY OF QUASARS

In 1963, Dutch-American astronomer Maarten Schmidt was puzzling over a "star" called 3C 273 that gave out light and radio waves. When he analyzed the light, it didn't make sense. Then he realized: 3C 273 is farther away than most galaxies and the expansion of the Universe stretches the wavelength of its light. 3C 273 was called a "quasi-stellar radio source", soon shortened to "quasar".

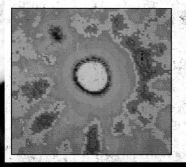

The first of thousands of quasars to be discovered, 3C 273 is imaged here by X-rays. It is one of the nearest quasars, just 2 billion light years from us. The average quasar is no bigger than our Solar System, yet brighter than a trillion Suns.

How massive black holes were born

Supermassive black holes, and their surrounding quasars, are probably a natural by-product of the birth of galaxies. New research reveals that most quasars were born at the same time – about two-and-a-half billion years after the Big Bang that created the Universe.

A cross section of the accretion disc reveals its bowl shape, moulded by the forces of gravity, rotation, and the tendency of hot gas to expand

As gas in the accretion disc moves inwards, it gets hotter – and glows brighter. This glaring gas gives quasars their brilliance

Stars are pulled closer to the black hole and ripped apart by its gravity

A reservoir of dark gas and dust clouds lies at the outer edge, and will gradually be swallowed up by the black hole. In a radio galaxy, this dust obscures the brilliant core

Soon after the Big Bang, gas clouds came together under the pull of gravity and merged to form galaxies

At the centre of a young galaxy, gravity pulled the stars and gas clouds together, eventually creating an enormous black hole

As matter continued to fall into the hole, it was violently heated – and an active galaxy was born

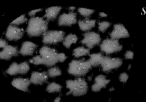

BLACK HOLES NEED FOOD

For all its awesome might, the accretion disc is not a permanent feature. Eventually, all its gas and dust will spiral into the black hole. Deprived of its food supply, the black hole lurks starved and unsuspected, and the surrounding galaxy looks entirely normal. But if more stars and gas come too close, the hole "wakes up" and feeds voraciously: the quasar is rekindled.

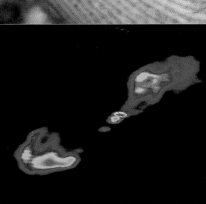

Imaged by a radio telescope, the radio galaxy Centaurus A displays two jets blowing huge bubbles in space.

Mirages and ripples

WHEN IT COMES TO TRACKING DOWN BLACK HOLES, astronomers are very resourceful. To find something so dark, they have to seek out telltale clues. Streams of snatched hot gas from an orbiting companion star, disruption in the hearts of galaxies and quasars, bursts of gamma rays – all these may be signs of a black hole at work. Recently, astronomers have discovered two new clues. Both rely on gravity and the effects predicted by Einstein. The first involves seeking out cosmic mirages. When light from a distant object passes close to a region of strong gravity, it is bent or distorted, giving rise to some bizarre optical effects. The second method entails detecting "ripples in space" – gravitational waves – created by the movement of massive objects.

Cosmic mirages

Einstein's theory of relativity says that when light passes close to a massive object such as the Sun, it is bent (see p. 18). A black hole has such strong gravity that it deflects and focuses light from a distant star so that it appears brighter than normal.

Gravitational waves may be caused by two massive bodies – black holes or neutron stars – in orbit about each other. They may also be produced by a supernova explosion or the merger of two neutron stars

The waves – "ripples in space" – spread out from their source at the speed of light

Light from a star in the Large Magellanic Cloud sets out on a journey of 170,000 light years towards the Earth

The light rays enter the galactic halo – a zone of scattered old stars surrounding the Milky Way

In the halo, the starlight passes near a black hole, where the warped space focuses the light rays

GALACTIC MIRAGES

A black hole works like a small lens, and its main effect is to make a star behind it look brighter. The distortion of space close to a galaxy, on the other hand, acts like a giant lens, spreading the light from a distant object so that there appear to be several objects instead of one. Here, a galaxy 400 million light years away deflects and splits light from a quasar 8 billion light years behind it, creating four quasar images around the central galaxy.

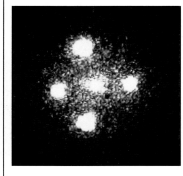

Four mirages of the quasar 2237 + 0305, popularly known as the Einstein Cross, captured by the Hubble Space Telescope.

Machos at work

The warped space near a black hole works like a lens in a telescope, bending and focusing the light rays from a distant star so that it temporarily appears brighter. Astronomers have observed this effect in stars belonging to the Large Magellanic Cloud, a small galaxy orbiting the Milky Way. They believe that the culprits are black holes or neutron stars living in the outer regions, or "halo", of our Galaxy. These have been nicknamed MACHOs – MAssive Compact Halo Objects.

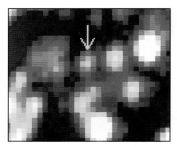

A star in the Large Magellanic Cloud at its normal brightness.

Ripples in space

If both members of a double-star system are supermassive giants that end their days exploding as supernovas, the end result may be a double black hole system. Like two circling speedboats, the black holes orbit around each other creating massive wakes of gravitational waves that spread, like ripples, through space.

Bouncing laser beams should be able to detect microscopic changes in the distance between two mirrors several kilometres apart as a gravitational wave passes through. This is a small prototype to test the laser systems.

As the black holes radiate gravitational waves into space, they spiral closer and closer together. This makes them emit waves even faster and more furiously. Eventually, they merge

WAVES THAT STRETCH AND SHRINK

Gravitational waves are shudders in the fabric of space, caused by the movement of massive bodies. Einstein predicted that the waves should exist, but no one yet has found them. In theory, they should shrink or stretch any object they pass through. But the effects are very small. A gravitational wave passing through an iron bar 1 metre (39 inches) long would change its length by the diameter of the nucleus of an atom. Detectors need to be extremely sensitive.

Gravitational waves cause distortions in the shape of space, just as waves ripple the surface of the sea

The most solid of objects are distorted when a gravitational wave passes through them. But the amount of stretching or compression is tiny

At the Laser Interferometer Gravitational Wave Observatory in California, laser beams travelling through tunnels 4 km (2.5 miles) long should register the tiny changes produced by gravitational waves

BRIEF BRIGHTENING

In 1993, astronomers found a star that suddenly brightened in the Large Magellanic Cloud. After a month, it faded again. The best explanation is that a black hole passed in front and briefly focused the star's light our way. The hole weighed only one-tenth as much as the Sun, and lay 20,000 light years away.

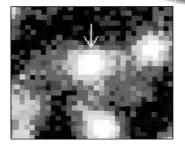

Star brightens as a black hole passes in front, focusing its light.

An optical telescope detects the brightening of a star in the Large Magellanic Cloud

Biggest black hole

A VERY DARING IDEA HAS EMERGED IN RECENT YEARS: we may be living within a black hole! Some scientists have suggested that the entire Universe is a huge black hole, of a rather different kind. It is not surrounded by an event horizon, but curves back on itself like the surface of a balloon. The result is the same: you cannot escape. The Universe has no central singularity. Instead, it had a singularity in the past, the Big Bang in which the Universe began – and it may collapse back into a singularity in the future, the Big Crunch. The theory links black holes and universes so closely that it predicts a black hole can actually create a new universe.

TIME

The Universe keeps expanding, pushing the galaxies away from each other as the space between them is increased

As time passes and the Universe cools down and expands, quasar activity dies down, leaving normal galaxies

A couple of billion years after the Big Bang, the Universe is full of quasars

THE BEGINNING
No one knows what caused the Big Bang: it may have been a fluctuation out of literally nothing. Fractions of a second afterwards, the temperature and density in the cosmic fireball were almost infinite. In the inferno, dozens of strange particles were created which, as the Universe cooled and expanded, formed the nuclei of the first atoms. Around 2 billion years later, these clumped together to make quasars – violent young galaxies with black holes in their hearts. Meanwhile, the Universe continued its expansion.

The Universe today, 15 billion years after the Big Bang, has expanded until it consists of widely separated galaxies

Universe 1

When the Universe is about 50 million times its present age, expansion will stop – if there is enough matter to counteract the momentum created by the Big Bang

Universe 1

Born to a black hole
Far from being merely cosmic sinks, black holes may give birth to other universes. "Baby" universes may "bud" off black holes to grow as universes in completely different dimensions with totally different properties. Depending on the amount of matter in each universe, they will be of different sizes but, like ours, will expand and contract, and produce still more universes along the way.

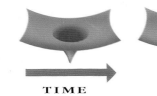

TIME

The remains of a supernova in our Universe (universe 1) starts to collapse and form a black hole in the usual way

A baby universe (2) buds off from the black hole. It detaches into a different dimension and becomes independent. This is the Big Bang for universe 2

Universe 2

Universe 2 starts to expand. Black holes that form inside it will create sites where other universes can bud off

From big bang to big crunch

If our Universe is a black hole, then it has a finite, and predictable, lifetime. At the Big Bang, the Universe starts expanding – shown here (not to scale) as a series of inflating balloons. The galaxies and other constituents of the Universe lie on the skin of the balloon, which represents space and carries the galaxies apart as it swells. The history of the Universe is shown on successive strips of the balloons. The Universe expands until it reaches its maximum size; then gravity wins over the momentum from the Big Bang. The Universe begins to contract. The galaxies move closer together and collapse into another singularity – the Big Crunch.

DARK MATTER APPLIES THE GRAVITATIONAL BRAKES

The fate of the Universe depends crucially on how much matter it contains. Too little, and there will not be enough to exert the gravitational pull needed to "brake" the growth of the Universe; it will expand forever. With sufficient matter, it will recollapse. Adding up all the visible matter in the Universe gives only 10 per cent of the mass needed to apply the brakes. But astronomers have evidence that the Universe contains huge quantities of

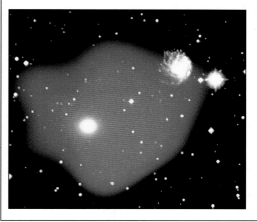

"dark matter". Perhaps 90 per cent of it consists of this invisible matter, which could comprise exotic subatomic particles or even vast numbers of black holes.

The NGC 2300 cluster of galaxies is embedded in a gas cloud (coloured magenta) weighing 500 billion Suns. The gravity from a vast amount of dark matter must be holding the gas cloud together.

Galaxies crash together. Temperature of whole Universe rises

Maybe our Universe will bounce back in another Big Bang to create a new – and very different – universe

IN THE END

After perhaps a million million million years, gravity will draw everything in again – the Big Crunch. About a year before, galaxies will collide, and the temperature of space will rise to higher than the surface of a star. An hour before, supermassive black holes in the centres of galaxies will merge and the Universe will dramatically collapse into a point of infinite density.

The Universe may continually oscillate between Big Bangs and Big Crunches

As the Universe shrinks, the galaxies move closer and closer together

Gravity asserts itself, and the Universe starts to collapse

TIME

Universe 1

Universe 1 collapses in a Big Crunch; universe 5 buds off

A black hole in 5 gives rise to universe 6

Universe 2 buds off from a black hole in universe 1

Universe 7 is spawned by a Big Crunch in 5

Universe 4 buds off when universe 2 collapses in a Big Crunch

A black hole in 2 creates universe 3

Our Universe may give rise to a whole series of universes

Deliberately created

Black holes and Big Crunches can both provide "budding sites" where new universes can start. From one universe, they can spawn a network of independent universes. Although each baby universe is different, it will inherit some of its parent's "genes". This may be how our Universe came to be. American cosmologist Ed Harrison even suggests that our Universe is so ideally suited for the development of intelligent life that it may have been created by an advanced civilization in another universe. Perhaps the scientists in that universe reached the stage of making black holes in the lab, and one budded off to become our Universe. We may be just the result of someone's experiment!

Glossary

ACCRETION DISC Disc of fiercely hot, glowing matter spiralling into a black hole.

ACTIVE GALAXY Galaxy undergoing a violent outburst in its central regions; see *blazar, quasar,* and *radio galaxy.*

ANNIHILATION Total destruction of matter in a burst of energy, for example when it meets *antimatter.*

ANTIGRAVITY A gravitational field that repels, rather than attracts, matter and light rays.

ANTIMATTER The exact opposite of matter. When matter meets the tiny amount of antimatter in the Universe, the two *annihilate* each other.

ANTIPARTICLE Atomic particle that has exactly the opposite properties of its matter counterpart; for instance, a *positron* is an antimatter *electron.*

ATOM Smallest part of an element, such as hydrogen, oxygen, or carbon, that can take part in a chemical reaction. Most of the *mass* of an atom is concentrated in its nucleus, which measures about a millionth of a millionth of a centimetre across.

BIG BANG The violent event that gave birth to our Universe, some 15 billion years ago.

BIG CRUNCH The ultimate collapse of the Universe that may take place in the future if the Universe starts to contract.

BINARY SYSTEM See *double-star system.*

BLACK HOLE Collapsed object whose gravity is so strong that nothing – not even light – can escape it. As a result, the object is black; and it is a hole because things that fall "in" can never escape.

BLAZAR Type of *active galaxy* angled in such a way to us that we look almost directly onto its *accretion disc* and jet.

BLUE SHIFT Shift in the *wavelength* of radiation emitted by an object when it is approaching us. The *Doppler shift* makes the

wavefronts bunch closer together, causing the light to appear of a shorter wavelength, and hence bluer.

COSMIC CENSOR Mythical being who dictates that *singularities* must be surrounded by an *event horizon.*

CORE In a star, the central region that is undergoing *nuclear fusion*; in a galaxy, the innermost few *light years.*

DARK MATTER Invisible matter that is believed to make up 99 per cent of the mass of the Universe. It may be in the form of black holes, or possibly exotic particles.

DENSITY Degree of "solidity" of a body: its *mass* divided by its volume.

DOPPLER EFFECT Change in the observed frequency of sound or radiation that takes place when the observer and the source are moving relative to each other.

DOUBLE-STAR SYSTEM System of two stars in orbit about each other.

DUST Microscopic grains in space that absorb starlight. The grains are "soot" left by dying stars, and they sometimes clump together in huge dark clouds.

EINSTEIN-ROSEN BRIDGE "Throat" of a black hole in one universe connecting up with one in a different universe. In theory, it is a bridge from one universe to another.

ELECTROMAGNETIC RADIATION Radiation made up of magnetic and electrical fields that move at the speed of light. It ranges from radio waves (long *wavelengths*), through light to *gamma rays* (very short *wavelengths*).

ELECTRON Tiny particle with a negative charge, often in orbit about the nucleus of an *atom.*

ERGOSPHERE Region surrounding a spinning black hole, between the *static limit* and the *outer event horizon*, in which it is impossible to be at rest.

ESCAPE VELOCITY Speed a body needs to travel in order to escape

the surface *gravity* of a star or planet. Escape velocity depends on size as well as *mass*; the smaller the object, the higher the escape velocity.

EVENT HORIZON The "edge" of a black hole: an imaginary surface where the *escape velocity* reaches the speed of light. It is situated at the *Schwarzschild radius.*

GAMMA RAYS The highest energy, shortest wavelength *electromagnetic radiation* of all. Theory predicts that it should be emitted when *mini black holes* explode.

GENERAL RELATIVITY Theory of relativity that describes how matter behaves in the presence of strong gravitational fields.

GRAVITATIONAL LENS Distortion of an image – or the production of many images – by a powerful gravitational field.

GRAVITATIONAL WAVES Ripples in space, which travel at the speed of light, produced by the movement of very massive bodies.

GRAVITY Force of attraction that is felt between bodies, such as the pull between the Earth and the Moon.

INFRARED Heat radiation, intermediate in *wavelength* between light and radio waves.

LAST STABLE ORBIT Closest an object can circle a black hole without being pulled in.

LIGHT YEAR Distance covered by a ray of light travelling at 300,000 km/s (186,000 miles/s) in a year. It is about 9.5 million million km (5.9 million million miles).

MASS Amount of matter making up a body. On Earth, the mass of a body is equal to its weight.

MINI BLACK HOLE One of many tiny black holes, with the *mass* of a mountain but the size of an *atom*, that are believed to have been created in the *Big Bang.*

NAKED SINGULARITY *Singularity* not surrounded by an *event horizon.*

NEUTRINO Minuscule particle with little or no *mass* and no charge that travels at the speed of light.

NEUTRON Electrically neutral particle that makes up part of the nucleus of an *atom.*

NEUTRON STAR Collapsed star composed mainly of *neutrons.* Pulsars are young, fast-spinning neutron stars.

NUCLEAR FUSION Nuclear reaction in which one kind of *atom* (for example hydrogen), under extreme heat and pressure, turns into another (for example helium). The energy released by fusion keeps stars shining.

PROTON Positively charged particle that forms part of the nucleus of an *atom.*

PULSAR See *neutron star.*

QUASAR Brilliant *core* of a distant young *active galaxy*, whose outer regions are often too faint to be visible.

RADIATION See *electromagnetic radiation.*

RADIO GALAXY *Active galaxy* that gives out as much energy in radio waves as it does in light. Most of the radio emission comes from two giant clouds ejected from the galaxy's *core.*

RADIO TELESCOPE Telescope that picks up radio waves from objects in space.

RED GIANT Old star whose outer layers have billowed out and cooled down.

RED SHIFT Shift in the light of a retreating object towards red *wavelengths*, caused by the *Doppler effect.*

RELATIVITY See *general relativity* and *special relativity.*

SCHWARZSCHILD RADIUS Radius of the *event horizon* around a black hole.

SINGULARITY The centre of a black hole; a point (or ring) of infinite density that occupies zero space.

SOLAR MASS Mass of the Sun; a "standard weight" against

Index

which other objects in the Universe can be compared.

SPACE-TIME Four-dimensional description of the Universe in which length, breadth, and height make up three dimensions, while time makes up the fourth.

SPAGHETTIFICATION Gravitational stretching of a body falling into a black hole.

SPECIAL RELATIVITY Branch of relativity dealing with the behaviour of objects travelling close to the speed of light.

STATIC LIMIT A limit close to a black hole inside of which it is impossible to remain at rest.

STELLAR MASS BLACK HOLE Black hole produced by the explosion of a massive star as a *supernova*. Most weigh about 10 *solar masses*.

SUPERMASSIVE BLACK HOLE Black hole located at the centre of a galaxy. These holes, formed by material falling onto the galaxy's *core*, may weigh billions of *solar masses*.

SUPERNOVA Explosion of a massive star at the end of its life.

WAVELENGTH Distance between wavecrests on any train of *electromagnetic radiation*. Short wavelength radiation (X-rays, for example) is more energetic than long wavelength radiation (such as radio waves).

WHITE DWARF Collapsed core of a normal star such as the Sun after it has lost its outer layers.

WHITE HOLE Exact opposite of a black hole; an object that spews out matter and energy.

WORMHOLE Object with two mouths in different parts of our Universe connected by a tunnel that allows two-way traffic. Wormholes may be safe shortcuts through space.

X-RAY SOURCE Region of extremely hot gas. Matter torn away from a normal star by a black hole or *neutron star* becomes violently heated and emits X-rays.

Acknowledgements

California Institute of Technology 12BR; /Bob Paz 39TR; **Camera Press**/Erma 35TL; **Hencoup Enterprises** 35CR; **Image Select** 18TR; **Max-Planck-Institut Für Quantenoptik** 41TR; **Faculty Files, Princeton University Archives** (used with permission Princeton University Libary) 21BC; **Ronald Grant Archive**/Castle Premier/Interscope Communications/Soissons-Murphey Productions/De Laurentis Film Partners ("Bill & Ted's Excellent Adventure") 32TR; **Science Photo Library**/Dr C. Alcock, MACHO Collaboration 40 BR, 41BC; /J-L Charmet 16TR; /T. Craddock 36TR;

/Hencoup Enterprises 13TL; /A. Howarth 24CL; /Lawrence Berkeley Laboratory 34BL; /W. & D. McIntyre 25TL; /Max-Planck-Institut Für Extraterrestrische Physik 14CL; /A. Morton/D. Millon 36CL; /NASA 12TR, 16BC, 16BR, 43CA; /NASA/Space Telescope Science Institute 10BL, 38BL, 40 BL, endpapers; /NRAO/AUI 8BL, 37CR, 37BR, 39BL; /NRAO/F. Yusef-Zadeh 37BL; /D. Parker 15BR; /Royal Observatory, Edinburgh/AATB 10TR; /Rev. R. Royer 18CL; /F. Sauze 22TR; /Smithsonian Institution 13CL; /X-Ray Astronomy Group, Leicester University 39CRA; **ZEFA** 23BR, 25CLA.